Pushes and Pulls

by Linda Ward Beech

This boy moves his boat by pushing
the pole against the ground.

Contents

Things That Move

Biker

How do things move?

How does a boat move in the water?

How can a biker ride in the park?

How does a tiger run through
the grass?

Tiger

Chapter 1

Big Idea Question

What Are Pushes and Pulls?

Things do not start or stop moving by themselves. They need a **force** to make them move. A force is a **push** or a **pull**.

Pushes and Pulls

What is a push? A push can move something away.

This man is pushing a cart.

What is a pull? A pull can move one thing toward another.

This boy is pulling a wagon.

Start Moving

A push or pull can make something start to move. How did the water skier start to move? A boat pulled him. The pull is a force.

When an object is moving, it is in **motion.** This horse is in motion. It pushed off the ground to jump.

Change Direction

Pushes and pulls can also make moving things change **direction**. Direction is the path an object takes. The player kicks the ball to change its direction. The kick is a push.

The man pulls the string back toward him. The pull will change the yo-yo's direction.

Stop Moving

Pulls can slow down moving things. The dogs will stop when the person pulls on their leashes.

Pushes can also make things stop. The players push to stop one another.

Find the Pushes and Pulls

Forces make things move. A push or pull is a force. Where are the pushes and pulls in this picture? Two are labeled to get you started.

pull

push

Big Idea Question

What Ways Do Objects Move?

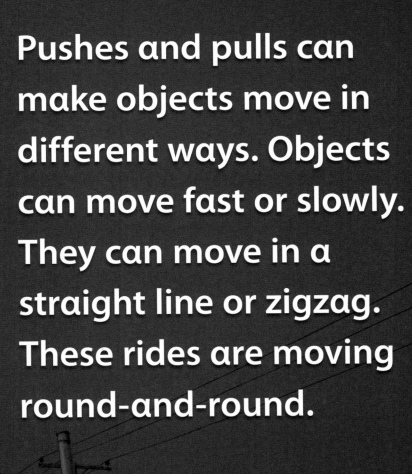

Pushes and pulls can make objects move in different ways. Objects can move fast or slowly. They can move in a straight line or zigzag. These rides are moving round-and-round.

Fast and Slow

When a dog is moving, the dog is in **motion**. Forces make things move fast. Forces make things move slowly.

This dog is running. It is moving fast.

This dog is walking. It is moving slowly.

In a Straight Line

This boy is pushing a mower. The mower moves in the direction of the push. It moves in a straight line.

Falling

Gravity is a force that pulls all things toward Earth without touching them. Falling downward is another way objects move in a straight line.

The ground is covered with oranges that fell from these trees.

Zigzag

These skiers move on the slippery snow. They zigzag down the mountain.

Back-and-Forth

These children pull on both ends of a rope. The rope moves one way. Then it moves the other way. The children pull the rope back-and-forth in a straight line.

This tennis ball is moving back-and-forth across the court. The ball changes direction each time a player hits it.

Vibration

When you pluck a rubber band, it vibrates. When something vibrates, it moves back-and-forth.

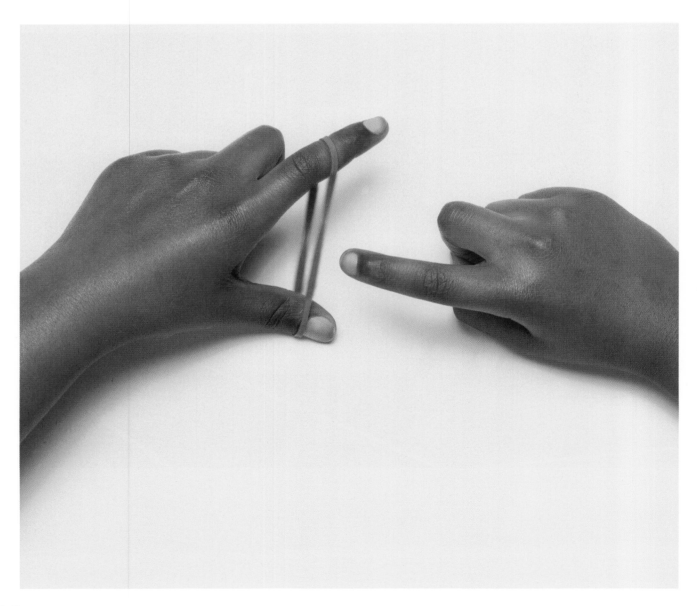

This person strums the strings on a guitar. You can tell which strings are vibrating. The strings are moving back-and-forth so quickly that they are blurred.

vibrating strings

Round-and-Round

Objects can also move in a circle. This dancer is moving. She pushes herself round-and-round.

This player balances a basketball on his finger. He gives it a push. The ball spins round-and-round.

Chapter 3

Big Idea Question

How Do Magnets Pull Objects?

This machine uses a **magnet** to pick up scrap metal. Magnets produce a force. The force can attract, or pull, some metals, such as iron.

Magnets Move Objects

Magnets can pull objects. They can pick up and hold objects made of some metals. Magnets come in many shapes and sizes.

Paper clips and tacks have iron in them. Magnets pull them.

All magnets have two different sides called poles. One pole is the north pole. The other pole is the south pole. The poles are where a magnet's force is strongest.

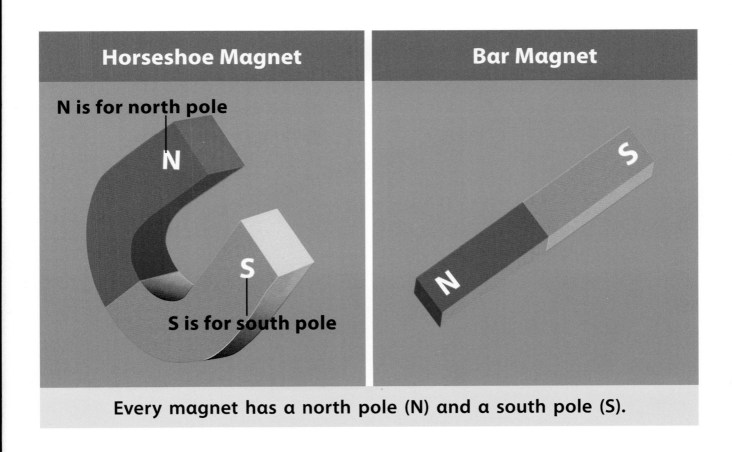

Every magnet has a north pole (N) and a south pole (S).

Poles that are different will pull toward each other. A north pole and a south pole pull two magnets together.

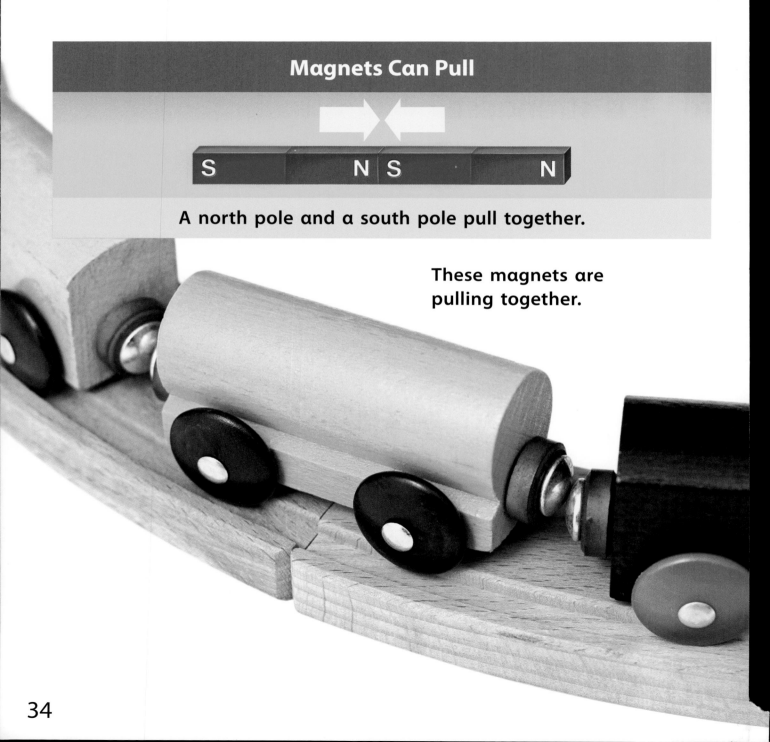

Magnets Can Pull

S   N S   N

A north pole and a south pole pull together.

These magnets are pulling together.

Poles that are the same will push away from each other. Two south poles push away from each other. So do two north poles.

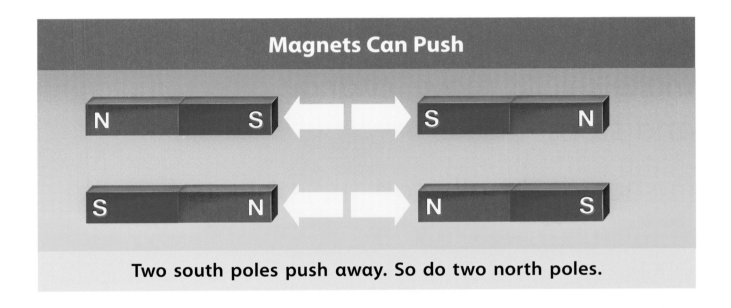

Magnets Can Push

N   S   →←   S   N

S   N   →←   N   S

Two south poles push away. So do two north poles.

Conclusion

Pushes and Pulls Move Objects

Pushes and pulls are forces. They can make things move, change direction, and stop.

Pushes and pulls can put objects in motion. They can make objects move in different ways.

Magnets can pull objects made of some metals.

Glossary

direction (page 12)
A **direction** is the path an object takes.

The ball changes **direction.**

force (page 7)
A **force** is a push or a pull.

The dogs use **force** to pull the man and sled.

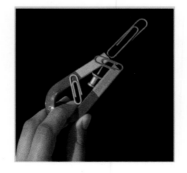

magnet (page 32)
A **magnet** is an object able to pull some metals toward itself.

This **magnet** pulls metal paper clips toward itself.

motion (page 11)

When an object is moving, it is in **motion.**

The horse is moving. It is in **motion.**

pull (page 9)

When you **pull** something, you move it toward you.

The boy **pulls** the wagon filled with groceries.

push (page 8)

When you **push** something, you move it away from you.

The man **pushes** a cart carrying trees.

Index

National Geographic School Publishing
Hampton-Brown
www.NGSP.com

Printed in the USA.
RR Donnelley, Johnson City, TN

ISBN: 978-0-7362-5547-9

10 11 12 13 14 15 16 17

10 9 8 7 6 5 4 3 2